Drones

The Ultimate Guide

By Casey Publishing

Table of Contents

Introduction

Drones – a word we all know given the press coverage these little flying robots have had in recent times. Not so long back, drones were a hobby for those who had plenty of money to potentially crash and burn with but now there are drones of all sizes and all prices, something to suit every pocket and every level. But, for those of you new to the drone, what exactly is it?

Drones have been about for many years in one format or another. Some of you will know them by their other name – a UAV, or Unmanned Aerial Vehicle and some of you will be well aware of their military history. Unmanned drones have long been used for reconnaissance and even for shooting down the enemy but, rather than talking about those, I want to talk to you about the recreational drone, the smaller version that you can buy just about anywhere now and use in and around your own home. In the next few years, drones are, quite literally, expected to take off and take over and later, I will tell you about some of the current and expected uses of these intelligent little robots. For now, let's take a look at what a drone is, where they came from and where they are headed.

There is no shame in asking what a drone is, no embarrassment because it isn't a daft question. The word, 'drone' covers a multitude of different things but, technically, a drone is an unmanned aerial vehicle that is capable of using GPS tracking to navigate by itself. In real terms, in today's world, a drone is a UAV that may or may not be capable of flying autonomously but is seen as less of a large military-type vehicle and more of a smart radio-controlled airplane or helicopter, more often called quadcopters.

A high percentage of these vehicles are equipped with cameras and some even sport systems that help them to avoid collisions and crashes. Many have highly technical settings that you would not normally find on your average radio-controlled helicopter, putting them firmly in the middle, somewhere between a traditional radio-controlled flying machine and a drone.

The Olden-Day Drones

I say olden-day but, really, I'm only going back a few years, back to when drones started to become popular. Getting one of these to stabilize was a lot of hard work, especially when you consider that some of the early ones were constructed of a frame of wood, gear assemblies and brushed motors and no less than three gyroscopes. LiPo batteries were unheard of then and you would be lucky if you managed to fly it for any more than a couple of minutes at a time.

10 years on and there have been some significant changes! Autonomous drones are everywhere, using GPS for navigation and fully stabilized. Everything needed, including the software and all the hardware, to build a drone, are now available anywhere and building your own

is so easy now it can be done in hours. Prices for parts have dropped significantly, especially the highly efficient processors, sensors, brushless motors and LiPO batteries. Drones are a major hobby now and it's fair to say that the leading manufacturer is DJI with its range of Phantoms and other UAVs. Off-the-shelf systems that are ready to fly are sparking no end of imaginative use and, over the next few years, we can fully expect to see more and more of them.

We need to give thanks to the developers and researchers who spend hours every day improving the technology and to the open-source community who collaborate and build this technology, making it available to all and not just the military! This technology has led to several businesses starting up, with Amazon kickstarting a revolution by announcing an aerial delivery system. Now we have businesses dedicated to component manufacture, developers for the software and the electronics, service providers, and a good deal more besides, all settling in for what is looking likely to be the long-term.

Now the world is looking at the drone, not just for the fun hobby side of it but for the more commercial side, applications that are likely to change how some industries do things but that is some way off in the future. First, we must go through the raft of startups that appear, some of which will fail and some of which will succeed with their technology. Weeding out the good from the bad will take time but, in the meantime, we can still play. We can still take our drones out and use them purely for fun, for photography, shooting videos, racing or just the sheer simplicity of flying one around your local area.

In the next chapter, we are going to start looking at how a drone flies and how the control system works.

Chapter 1: Flight of the Drone

To understand how a drone works, you must understand how they differ from the traditional remote-control airplanes and helicopters. There is one major factor that separates these from the drone and that is that the drone is, to a certain extent, autonomous, which means that they can maneuver and navigate without needing any pilot input. The drone can stabilize itself, can follow a GPS navigation system where the traditional helicopters require more than a small amount of input – that is what makes the drone an incredibly intelligent robotic machine.

Pull up a picture of a radio-control helicopter and a drone – there are obvious differences, the main being that a smart helicopter, or drone, is multi-rotor. Yes, you can get multi-rotor RC vehicles, but a drone requires these multiple rotors so that it is fully self-reliant. A multi-rotor setup offers many benefits, not least the fact that the more rotors you have, the more fail-safes you have. If one were to stop working, for example, your drone won't fall out of the sky and hit the ground with an expensive thud. The other motors will simply work together to keep it in the air, giving you time to bring it safely down. And, if there are several rotors on the drone, you get a good deal more lift which, in turn, means you can carry more weight with the drone, like a decent camera for example. Lastly, the more rotors, the smaller they can be and this, in turn, makes everything far easier to manage and safer to use.

Having all these rotors is wonderful but, to get them moving you are going to need some kind of power source. An off-the-shelf drone will usually come equipped with a battery that can be removed. These batteries typically allow around 10 to 12 minutes of flying before they need to be recharged but some will offer less, some more. You can purchase spare batteries from any drone suppliers or from multiple places on the internet and, provided your drone can take it, you can get bigger batteries that will give you around 20 to 25 minutes of flying time. However, be aware that the more powerful the batter is, the heavier it is and is this will affect flight time.

So, you have your drone with multiple rotors and you have your fully charged battery. To get this thing off of the ground and up in the air requires a controller that you use to launch your drone, move it around and land it again. There are lots of different controllers – some look a traditional RC controller, others look like game controllers. The latest drones can even be controlled by a smartphone or tablet. Really, it matters not what your controller looks like so long as it does the job and can communicate with that drone and it does that through the use of radio waves. A drone normally runs of 2.4 GHz waves, usually through a Wi-Fi system, which is why tablets and smartphones work so well as controllers because there is no need for any extras to use that radio spectrum.

The smartphone is also partly responsible for some of the technology found on some of the latest drones. Like the GPS chip, for example, this is used to send the controller its location, as well as logging the point of take-off should it need to come back to the controller, without any assistance. Another vital addition also found in a smartphone accelerometer is the gyroscope and, over time, these have shrunk in both size and price but grown in power. Because of these factors, these accelerometers are now more accessible and can be inserted into so many more gadgets, including the hobby drone.

So, when your drone is up in the air, it is the sensors that keep it there. One of those sensors is an altimeter, telling the drone at what altitude or height it is flying at. If for example, you set your drone, so it hovers in one place, the altimeter will pass the information on to the drone to stay at that height. The GPS sensor is used to keep the drone inside of the set of X and Z axes, changing the course of the drone when it gets caught by the wind. Many drones, especially the newer, more complex models can stay on a straight course in gusts of wind at up to 50 mph making them much easier to fly than ever before.

Flying is one part of it; the other is landing it. Most drones will, because of their programming, carry out controlled landings, slow and steady, pretty essential for an aircraft with propellers. If you try to land one too quickly you could end up in a vortex ring state, something that is known by most experienced helicopter pilots. It may not have a name in the world of drones but, if you drop the drone too quick out of the sky, it will get caught up in the vacuum caused by its propellers, and this is not very easy to get out of it, n matter how much experience you have. Why? Because the temptation to is to push hard on the throttle to try and push your way out of it and that just increases the strength of the vacuum, pulling the drone down faster and harder.

One last feature that can help your drone to stay on course is the camera. Some drones are able to use the camera as a kind of positioning system that helps it to land. It does this by making a map of what is below the drone and draws up a kind of grid that has points of safety on it. Should the drone drift, the camera can pull it back on course and sensors can work out how near to the ground the drone it. It is this technology that can make it look so easy to land a drone, but it still depends, to a certain extent, on you, the person with the controller in their hand.

Next, we look in deeper detail at what makes up a drone.

Chapter 2: Anatomy of a Drone

Up until now, I've taken it a bit easy on you but now we get to the real nuts and bolts if you will pardon the pun, of what makes up a drone. Every single part of the drone is vital, and each plays its own part. Also, knowing what your drone is made up of and what each art will help you to understand which parts need to be inspected and when, as well as knowing which parts which can easily be upgraded and/or replaced. And, if you know your parts, you should be able to work out what any problems are that arise.

Overview – Drone Components

• Standard or Tractor Propeller

These are the propellers that you will find to the front of your drone and the reason they have the name of tractor props is that they will pull the drone like a tractor does but through the air. You may be looking at your drone in some confusion – most look the same no matter which angle you look at it from but trust me on this – there is a front and there is a back!

Drone propellers tend to be manufactured from plastic or, on the higher end models, carbon fiber, and this is to make them as light as possible. When you first start flying your drone, you can fit guards to the propellers to protect them and to protect anyone you may accidentally hit!

TIP – before you fly your drone – every time – inspect the propellers and always have a spare set with you. You should never, ever fly a drone that has damaged propellers on it.

- **Pusher Propellers**

These are featured at the rear of the drone and, as you would guess from the name, they push the drone forwards through the air. These also go by the name of "contra-rotation" propellers and this is because they will cancel out any torque on the motor when the drone is on a stationary level. Again, these may be constructed of plastic or carbon fiber and guards are available for them.

TIP - as with the front props, inspect them before you fly and do not fly with damaged pusher props.

- **Brushless Motor**

Pretty much any drone you buy off the shelf today will have a brushless motor, an out-runner motor which is far more reliable, far quieter and way more efficient than a standard brushless motor and they are usually electric. The design of the motor on your drone is very important because the more efficient they are, the more battery life you get and the longer you can fly. Drone motors are an ever-evolving component and, as each new iteration of the drone is released, we can expect to see better and more efficient motors on them.

TIP - Make regular checks on the motors and ensure they are kept clean of any dust and debris. Learn how your drone should sound – most of this is down to the motors – and then you will be able to hear when there is a potential problem. If the motor doesn't sound right, examine it. Fly the drone a foot or two off floor level near to you so you can see if there is any failure on a motor and always have spares to hand.

- **Motor Mounts**

On some drones, you may find that you have a motor mount built into the frame or fitted with landing struts. These are purely to mount the motors on and hold them steady.

TIP – examine the mounts regularly and look for any cracks that may be due to stress. If you find any and your drone is still covered by the warranty, get them fixed. If not, you will need to replace them – some manufacturers supply strengthener mounts. Also, when you first get your drone, look closely at the screws holding the mounts in. On occasion, manufacturers wind these in too tight and this can cause cracks in the frame. Again, send it back for repair or replacement – hairline cracks don't get better!

- **Landing Gear**

Some drones, especially those which require high clearance from the ground, may have skids, similar to those on a helicopter, mounted on the body but drones they are not designed to carry anything beneath them may not have any landing gear at all. You may also have a retractable landing gear on yours.

TIP - Examine the landing gear regularly, in particular, of your drone has a bit of a hard landing. This gear is what protects all the sensors, the camera, even the drone itself so it needs to be in tip-top condition. If you have the option, try to get a drone that has the retractable gear – if you are into photography and videos, this will keep the legs out of the shot. Again, keep them clean so the gear does not seize up.

- **Boom**

In some drones, the boom will be a part of the actual drone body while others will have a separate boom. These are meant to be tough as they need to survive a crash while not getting in the way of the downdraft from the propellers. A short boom makes your drone far more maneuverable while a longer one gives better stability.

TIP – examine the boom regularly – if it gets bent it will need to be replaced as it will interfere with flying.

• **Drone Body**

The drone body is the central part of the drone where the booms come out. The body is where you will find the sensors, the camera, processor, circuit boards and the battery – the life of the drone!

TIP – Unless you specifically purchase a waterproof version, your drone most likely won't like a fall into the water. It is very important that the main body is not submerged in water. Also, while a crash landing or a hard drop may not destroy the drone body, it may shake up what's inside so try not to bring your drone down too hard. If your drone is under warranty, do NOT open the body and tinker with what's inside as your warranty may be invalidated.

- **ESC – the Electronic Speed Controllers**

The ESC on a drone is a circuit that controls variable speed on the motor, controls the direction and, in some cases, may act as a brake. It will DC power into AC 3-phase for the brushless motor to run. These are a vital part of your drone, offering a compact yet highly powerful AC power pack. You don't need to do anything with these as they are housed in the main drone frame.

- **Flight Controller**

This is what reads the input from the sensors, the GPS, the battery and the receiver and regulates the speed of the motors to ensure that you can steer your drone. It also has a hand in controlling other autonomous systems on the drone, including Follow Me (if available), waypoints, failsafe, etc. The flight controller is what makes your drone function as it should. You do not need to touch or examine this in any way.

- **GPS**

The GPS sensor on your drone is a requirement for most autonomous modes of the drone, such as navigating by waypoints. The sensor will often have a magnetometer included so that it can provide compass headings, elevation, longitude, and latitude. If your drone doesn't have GPS, then you are very limited in how you can use it. Many of the most recent drones will also have Glonass and, together with GPS, your drone has many more satellites to use for positioning and it ensures more accurate flying and safety because your drone is unlikely to lose a connection to the satellites.

TIP – most up to date drones have the facility to let you set a failsafe point for home, which will bring the drone back to that point if the connection between it and the controller gets broken. However, you will likely need to have a certain minimum number of satellites first so make sure you buy a drone that has Glonass and GPS, so you get the most satellites.

- **Receiver**

Most likely, your drone will have a standard receiver unity, like that found with most radio-control machines. You need at least a quad-band receiver although 5 bands are generally recommended. If yours goes wrong, there are plenty to choose from on the market these days.

- **Antenna**

Your receiver will have an antenna and, depending on which type it is, the antenna will be helical or loose wire whip antenna.

TIP – these are easy to change, and you can upgrade the one you have if you want.

- **Battery**

LiPo or Lithium Polymer batteries are the best for a drone because they have a great combination of market life, energy, and power.

TIP - Carry spares with you otherwise you will be packing up and going home after a few minutes! Your drone will already have a battery, but you may be able to upgrade to a larger one that will have a longer flying time. Also, make sure you follow the storage and charging instructions for the batteries. Make sure you completely discharge your battery and then recharge it every now and again. Make sure the battery isn't overheating and never, ever, use a battery that has been damaged.

• **Battery Monitor**

A battery monitor is used to provide the flight controller with monitoring of rate power level while the drone is in the air. Your battery is the single most important aspect of safe flight – If you don't call your drone back before the battery runs out, it will crash or carry out an emergency landing.

TIP – learn how long you can fly on each battery, learn how to read the battery levels. Never push your drone beyond those limits; a rule of thumb is to bring it back when your battery level falls to 20%.

- **Gimbal**

The gimbal is a mount that pivots around the axes to give your drone stabilization and for pointing sensors and cameras.

TIP – If you intend to use a camera for stills and video, then you need a good gimbal. You can have the best camera in the world, but it will all be for nothing if your gimbal is of inferior quality. Also, if you are getting a lot of vibration from the drone to the camera, think about changing out or, if you don't have any, installing dampers on the gimbal.

- **Camera**

Choose a good action camera, such as a Go Pro or other similar HD unit and make sure it has a decent amount of storage space or takes a good size of SD card. Some drones will allow for streaming in real-time while others simply recordto memory for playback. It doesn't matter how good you think you are at photography, if your camera isn't up to scratch, you won't get good shots.

TIP – try to match the gimbal to your camera for the best results. Some gimbals are compatible with specific cameras, such as the Go Pro Hero cameras. Of course, you may have a drone that has the camera and gimbal built in, in which case, try to choose the same if yours need replacing.

Always buy the very best that your pocket allows for and, if you can get a drone with the integrated system, buy it as these are tested and proven to work.

- **Sensors**

We can use our drones for a good deal more than videography. Some of the more upmarket drones also have thermal sensors, along with others, fitted and can now be used in many different ways. We will cover some of these ways in a later chapter.

- **Collision Avoidance**

Today's drones have 2 different sensor types – the first is the thermal and Lidar sensors and the second are collision avoidance sensors, integrated onboard and these use several systems including vision, infrared, lidar, monocular and Time of Flight sensors, to avoid colliding with objects.

That pretty much covers the major components that make up your drone. Next, we are going to look at how to choose a drone that suits you.

Chapter 3: Choosing the Right Drone

It's one thing to say that you are going to buy a drone but quite another to get the right one! It really isn't a case of just picking the first one you see; there are several things you need to look at and consider first. And you are going to be somewhat overwhelmed by the sheer choice from which to make your selection.

There are big drones and small ones, cheap right up to eye-wateringly expensive drones and so many features that your head will start spinning faster than the drone propellers. There are things to consider before you make your choice, especially your requirements and, more importantly, your abilities so cast your eye over these tips and think carefully before you make your purchase.

- **Your Skill Level**

This is the most important consideration. If you are a complete novice, have never flown any kind of drone before, then it's pointless buying the most expensive, all singing-all dancing model when you are likely to crash it on your first time out. Some drones are very easy to fly while others are hard. Much depends on their design and what features they are sporting.

Some can be flown by anyone, others require you to be something of an expert.

For your very first flight, go for something that isn't too expensive, is built fairly sturdily and has extra safety features, like guards on the propellers – you can buy these separately though. Some drones are aimed specifically at the novice market – they aren't the cheapest nor are they the most expensive, but they are far more intuitive and responsive. If, on the other hand, you already have experience under your belt then go for something that is a bit more advanced.

As long as you are honest about your skill level, you can narrow the range down.

- **The One That Fits What You Want to Do**

There are so many different things that we can do with drones so it's important to choose one that fits your requirements. For example, if you want to be able to fly yours outdoors then don't choose an indoor one. You need to one that is going to withstand wind gusts and indoor ones are simply not strong enough. If you are looking to do some trick flying, again, choose one that is able to do that, while if photography is your thing then there are quite a few that can take or already have a camera mounted on them.

That leads to another point – the camera will depend on the type of photography or videos that you want to shoot – cheap cameras with cheap gimbals won't give you professional results.

Decide exactly what you want to do with your drone and then look for a drone that can do it.

- **Decide Which Features You Want**

So, you've narrowed down your choice to suit your skill level and what you want to do with your drone so now you need to look at the features. There are loads of different features and each drone will have something specific, such as a camera. Now, while there are plenty of cameras, not all of them will be good cameras so if photography is your feature, make sure you pick one with a good camera and gimbal setup. Think about the type of controller you want, whether you want GPS or not, the size of the drone, battery life and so on. Even look at what accessories they come with as these might be just as important as features to some people.

Choosing the drone that suits you means choosing the one with the features that you want. For example, if you want to be able to start flying your drone immediately, pick one that is RTF enabled - Ready to Fly. On the other hand, one of the best ways to learn about your drone and what its components are is to purchase one that requires building first. If you are looking for durability, then choose one that has propeller guards or one that has the best reviews for surviving damage in a crash landing.

No matter which one you end up buying, just make sure it does what you want and has the features that you want

- **A Drone that is Decent Quality and Fits Your Cost Requirements**

Of course, the price is going to be a consideration, but this should be one of the last things to consider. Prices run form a couple of dollars into the thousands and knowing where to start is hard. Your budget will have some say in what you pay but narrow your choices down first using the criteria above. Once you know what you want in terms of skill and features, you can start to look at what those available to you cost. However, you should be able to balance affordability with the quality of the drone. For example, you might have two drones that meet your requirements but with two very different prices. Too cheap and you can pretty much guarantee that the materials won't be the best standard. Too expensive and it may well just be too much – the last thing you want is to destroy an expensive drone on your first flight by crashing it! And expensive doesn't always mean you are getting the best.

One of the first things to do once your choice is down to a few is to research each one and look for reviews. Talk to people in the drone community and see what they recommend.

Choosing a drone doesn't need to be difficult, you just need to do your homework first.

Chapter 4: How to Fly Your Drone

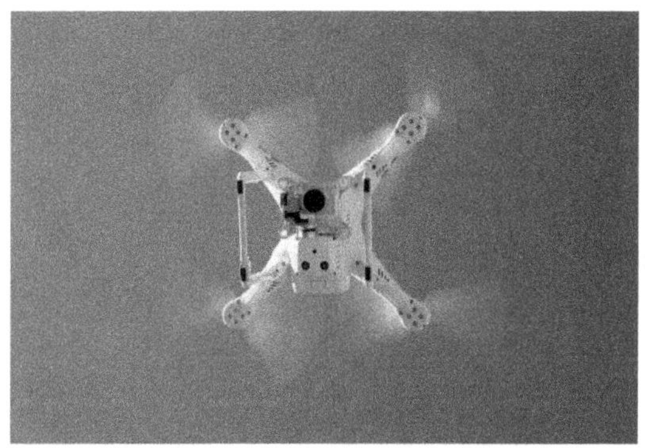

Flying your drone for the very first time is going to be quite a scary experience, more so if your drone was one of the expensive ones! You will be scared of crashing and this should make you be more cautious when your drone takes to the sky – at least, it should do! In this chapter, we are going to look at how to keep your drone in the air and out of the trees – or bushes – or lake, wherever it may crash land!

All drones are different in their capabilities, but the concepts of flying are pretty much the same. The most important thing to learn first is the terms that are used and what they all mean:

Terms and Definitions

- **FPV – first person view** – This is a system that lets you see what the camera attached to your drone can see, at the time it sees it. This is an incredibly useful feature if you intend to use your drone for photography or videography purposes. FPV also makes it much easier to fly your drone as you can better see where it is positioned and what direction it flies in.

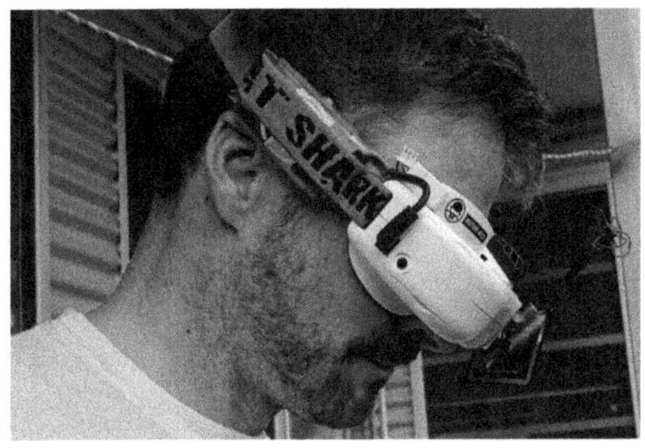

- **The line of Sight** - Get used to hearing this term if you take part in the drone community. All it means is that, during the flight, you can always see your drone. Many of the accidents you hear about tend to happen when the line of sight is lost.

- **Transmitter (remote control)** – This is what you use to control the drone, for navigation and communication purposes. There are loads of different transmitters on the market, some more advanced than others, while some are simply done via an app on a smartphone or tablet.

- **Camera** – Having an integrated camera on a drone is one of the biggest selling points and is one of the main reasons for using a drone. Cameras can take some incredible shots and videos, things that you could not have done before. If your drone does not have a camera, make sure it at least has a decent mount that will take a good camera.

- **Yaw** – This is a way of changing which direction the drone is flying in, using left rotation or right. This is normally done with the left transmitter stick although you can change this on certain transmitters. If you push right or left on the control for the yaw, the front of your drone will tilt in the requested direction
- **Throttle** – this controls altitude and is usually activated by pushing the throttle control down or up on the same control stick as the yaw.
- **Roll** – this is controlled by the opposite stick to the yaw and throttle, usually the right transmitter stick. Pushing the stick left or right will roll the drone in the chosen direction
- **Pitch** – the same stick, pushed forward or backward will control the

pitch, and the drone will tilt in the selected direction

- **Trim** – To increase the control you have over your drone, you use 'trim' to adjust how sensitive the yaw, throttle, roll, and pitch are. This function is usually used through a couple of buttons that are on the transmitter
- **Aileron** – relates to the right and left movement of the drone
- **Elevator** - relates to the forward and backward movement of the drone
- **Hover** – the ability to keep your drone stable and static in one position using good control of the throttle
- **Auto-Level** – putting the left and right transmitter sticks into a central position and allowing the drone level out

- **GPS Hold** – by centering these sticks, you can get the GPS position of the drone
- **Bank Turn** - by turning left or right you can move your drone in a circular movement

Learning How to Control Your Drone

The one thing that you can do to ensure full enjoyment of flying your drone is to learn the controls, learn how to maneuver it properly and this means understanding the remote control fully, how each part of it functions; this will take a little time but is an essential part of learning to fly your drone.

When you first pick up the transmitter, teach yourself to move the controls, in short, measured and smooth movements. Keep your drone just off the ground and in a fairly open space if possible – if it crashes, it won't hurt! By doing this, you can begin to learn how the stick movements affect the drone and how it moves.

As we mentioned above, the yaw, throttle, pitch, and roll are the main control functions and, as you get used to how they control the drone, you can make small adjustments in sensitivity, adapting your own style of flying. One of the main issues that new flyers face is that, when your drone is facing you, the sticks operate in the opposite way to when the drone is facing away. For this reason, you must keep your drone in sight all of the time or you could end up with the drone flying off into the sunset! This isn't too much of an issue if you have the fly home function that will bring your drone back – only if battery life will allow it though. However, you should not put all your faith into this, especially not to start with.

Expect to use the trim function a lot while you are flying your drone and learn how to use the left and right sticks to control the throttle, yaw, pitch, and roll.

Pre-Flight Checklist

So, you understand what your drone is made up of and have a basic understanding of the transmitter controls. However, before you take off on your maiden flight, there are some things you need to do first:

- Make sure that your drone batteries are charged up fully and that you have a couple of spares with you
- Ensure that the battery is inserted into the drone in the right way and is secured in
- Make sure that everything on the drone is tightened down and that there is nothing lose or hanging down anywhere
- Make sure that the satellite lock has been set and, if required, calibrated
- Make sure that the camera has been secured correctly, has an SD card in it if needed and is switched on
- Make sure the transmitter is switched on and working correctly

- Make sure that the propellers are not damaged, are fitted correctly and function as they should
- Make sure that you are in a clear area with no obvious obstacles and away from other people
- Make sure that the throttle stick is held in the downward position so that the drone cannot take off before you are ready
- Keep a safe distance from the drone before you launch it

Now you are ready for your first flight so, very slowly and very steadily, launch your drone with just a small amount of throttle and get ready for some fun. Just make sure that you always keep your drone in sight or you run the risk of losing control and the drone crashing.

Make your first few flights over grass; it is a much better cushion than concrete!

Flying the Drone

Ok, let's look at a few flight basics.

• Hovering your Drone

This is actually quite easy to do once you understand the yaw and the throttle. To hover your drone, begin with the drone on the ground and increase the throttle sensitivity just a little to get the drone off the ground and up into the air just a bit. When you get to the height you want, make a small adjustment on the right stick to prevent eh drone from turning or form rotating. Do this is very small movements. You may also need to move the left stick a little to get to the hovering position you want. If you can maintain your position and learn how to do this at other heights, you can get some great camera footage. When you take the drone higher, you will need to make good use of FPV to make your adjustments and keep the drone hovering.

- ## **Landing Your Drone**

There are a few ways to land the drone and crash landing isn't one of them! How you land is going to depend on which position the drone is in. If your drone is up high, you might find it better to just call the Return Home function and ease the throttle back, resulting in a smooth landing. If you are hovering, you will need to make some small decreases to the throttle to come back down slowly and steadily.

- ## **Flying in Different Directions**

Once you have got the hang of hovering you can move on to flying in different directions. To stay at the same elevation but to move about requires that you maintain control of the throttle. To go forward, push the right stick up gently and, to come backward, pull it back a little. To go right, move the right control stick to the right and to the left to turn the drone left. This might all sound very easy, but you won't find it easy until you have got the hang of it and this is where practice comes in When you first start to move in different directions, you will see that the drone will drop a little – don't panic, just gently push forward on the throttle and your drone will even out again.

- **Rotating Position**

One thing that many beginners get a little muddled with is the difference in rotating and flying left and right. In the last paragraph, I told you how to turn to the left or the right, a movement which will pitch the drone in the required direction and turn it. Rotating it is a different matter. To rotate your drone, use the left stick on the transmitter and move it the left or the right. One very good skill that you should practice is rotating while you are hovering – this will let you quickly change your direction in the air.

A Few Training Techniques

So, just to recap and add a few new ones, these are a few techniques that you should practice over and over to become a real pro at your flying your drone:

- Master hovering
- Master rotating in different directions while hovering
- Fly in circle and square patterns while keeping your drone completely under control
- Practice all of these at different altitudes

- Set 3 separate markers up in a wide-open space and then fly your drone between them, until you can do it almost with your eyes shut
- Now learn and master how to fly your drone in a figure of 8 and carry out bank turns to the left and to the right

Safety Advice

Yes, I know but drones aren't toys, not really and with inexperienced people they can actually be quite dangerous. You have propellers that are turning at high speeds and if you lose control of the drone, it can severely damage whatever or whoever it hits as it falls from the sky. And let's face it, when you are a novice flyer, crashes are par for the course so learning some basic safety can help to minimize the damage that occurs.

- Before you do any maintenance on your drone, make sure the battery has been removed
- Do not attempt to pick your drone up until the propellers have become completely still. They may only be plastic, but they can still inflict damage.
- If you lose control and your drone is falling from the sky and you can't regain any control over it, push the throttle so it is at the bottom. This way, the blades will not turn after a crash and your drone stands a much better chance of surviving the crash
- Never, ever fly your drone in crowded areas and make sure you know what the laws are regarding drones in your area – more about this later.

Emergency Procedure

Lastly, we need to look at what to do in the event of an emergency:

If you lose control and the drone flies off:

- If you can, make sure any person in the area is aware
- Try to keep your eyes on the drone for as long as possible
- Use CSC to stop the props in case the drone hits someone

If you lose line of sight:

- See if anyone else in the area can see it
- Try to increase the drone altitude if you can so you avoid hitting anything

- If you have FPS, use it to try and bring the drone back
- Use radar or maps to bring your drone home
- Use Return to Home if you can and pull the pitch stick back
- Use the yaw stick to turn the drone so its nose is facing the right way

If you lose radio contact:

- Keep your eyes on the drone and track it as far as possible to try to get its location
- If you can, get others to help you
- Realign the antenna if you can to try to get contact back
- Relocate the radio control transmitter to try to get contact back, move to higher ground if you can
- Engage Return to Home if you can

If you lose both visual and radio contact:

- Realign the antenna if you can to try to get radio contact back
- Relocate the transmitter to try to regain contact
- Engage Return to Home

If you lose GPS:

- Try to keep visual contact and bring it home manually
- Use the video stream to try to bring it home
- Use radar or maps to bring it home

Other

- If you lose your drone in a location that you know, do not attempt to recover it unless it is safe to do so. If the area is unsafe or not easily accessible, try to get help in recovering it
- Make sure you have a first aid kit to hand in case of injury to you or other persons
- Make sure you are aware of the location of the nearest medical facility if needed

In the next chapter, we will be looking at drone maintenance

Chapter 5: Maintaining Your Drone

Drones are a great deal of fun – until they go wrong. At that point, you have to work out what went wrong and how to fix it. Most problems on a drone can easily be solved by simply replacing a part and this is why it is so important to have a good stock of spare parts. Some of these will be cheap to buy, others won't but, considering the price you pay for your drone in the first place, it is well worth the expense of buying the spares than have to buy a new drone. We're going to talk about how to maintain your drone and which spares you should always keep in stock.

The most common spare parts

Some parts of your drone will need to be replaced more than others and propellers are just one of those parts. Trust me when I say that you will go through more propellers than you thought possible, especially when you are just starting out so always make sure that you have plenty in your bag. This next list is the spare parts that you will need to have on hand so that you can keep your drone in top working condition:

- **Propellers** – these break very easily so keep a good stock of spares

- **Gears** – these will wear out as time goes by and will need replacing
- **Rotor shafts** – like the propeller gears, these also wear out and, if your rotors get damaged, the shafts can also get damaged
- **Batteries** – battery life will die down and sooner or later, you will find that your battery doesn't last as long as it should. Keep a few spares so that you can't run out of power and also make sure you have a spare charger, so your batteries are always fully charged
- **Propeller guards** – by investing in guards, you can save yourself quite a bit of cash on replacing the propellers. Yes, the guards will break but they are cheaper to replace than the propellers so keep a few in stock

- **Memory cards** – if you are taking photos or videos there is nothing worse than running out of space for storing them on so have a few extra cards to hand
- **Wires** – those used inside a drone are not the thickest of wires and they will break. Keep plenty of wire to hand because you will need to use it to make repairs

As well as keeping a good stock of spare parts, you will also need specific tools for maintaining your drone, including:

- **A soldering iron** – you will be doing some soldering, but you don't need the most expensive solder iron. Just make sure that what you buy is fine-tipped as the parts are small and the wires are thin
- **Wire cutters** – preferably with wire strippers built-in

- **Double-sided foam tape or some Velcro** – used for mounting things on your drones, like transmitters or cameras
- **Heat-shrink tube** - this is useful for keeping external wires protected from being damaged which, in turn, cuts down on maintenance
- **Fiberglass and epoxy** – for repairing chips or cracks on the outer body casing

Do make sure that you purchase your spare parts from reputable sources, preferably those that deal in drone manufacture and parts. That way, you can be sure you are getting good quality parts and you won't waste time running about looking for things.

Maintaining Your Drone

To keep your drone in top condition and flying as it should be, you will need to do some things before you take it out to fly. Draw yourself up a checklist of maintenance and ensure that you do everything on the list.

Pre-Flight

- Make sure the propellers are not cracked, bent or broken. If they are, replace them
- Ensure that any moving part actually moves properly – each of these should be moveable by hand
- Make sure the drone bearings and any screws are tight
- Power the drone up and listen to it, make sure it sounds like it should

Each Week

- Check the shell for cracks and chips, repair any that you find
- Make sure that the shell and motor screws are tight
- Recharge or replace batteries as required
- Use alcohol wipes to clean the shell

Each Month

- Provided your warranty won't be affected, remove the shell and make sure the interior wiring is in good repair
- While you have the shell off, make sure the motor is OK

Professional Repairs

The more you fly, the easier you will find it to keep your drone maintained and in good repair but sometimes, you may come up with an issue that requires professional intervention. Trying to fix these things yourself could lead to worse problems so make sure you find a reliable professional repair outfit who can do the repairs for you. This will save you money, time and you will soon have your drone back in good flying order

Next, we are going to look at some of the uses of drones.

Chapter 6: Other Applications

There's more to a drone than just flying it around your yard; there are other things that you can do with your own drone and there are services that drones are now just starting to be used for.

Military

Drones have long been used for military purposes and, in the US alone, we can trace the use of UAVs back to 1917. The following are current and potential uses of drones for the future:

- **Bomb Detection**

Because drones are so small, they can get into spaces that other vehicles can't, and, with the added effectiveness of cameras, drones are the perfect vehicles for detecting bombs and potentially saving hundreds or thousands of lives.

- **Surveillance**

Most governments across the world use their defense services to carry out surveys on a regular basis. These are done to protect people and country and drones could be one of the more revolutionary ways of doing this. This would cut down on manpower and would provide a much bigger viewing field to survey as well as cutting down on interference to people as they go about their daily lives

- **Air Strikes**

UAVs have already been used for carrying out air strikes; former President of the United States, Barack Obama, has confirmed that they were used in Pakistan, attacking militants in the tribal areas. Drones can hover over areas of interest and can be operated by military personnel to carry out operations.

Non-Military

Drones are now being used more and more in some of these areas:

• Journalism and Films

The media world makes effective use of drones and many films now contain several scenes shot using a drone. This has revolutionized movies and some of the biggest names that we associate with this are Skyfall (James Bond), The Wolf of Wall Street (Leonardo di Caprio), Harry Potter and Game of Thrones, to name just a few.

In terms of journalism, drones can get to places that photographers can't, and we also see more and more aerial footage on news shows, in disaster zones, and so on.

- **Shipping/Delivery**

These are still in the pipeline but using drones for shipping and delivery purposes could be a revolutionary move for the world of shopping. The use of drones could see significant cuts in delivery times and could reduce the amount of manpower needed to get parcels out to recipients.

Amazon, the online marketplace giant, has already announced plans to use drones to deliver packages to wherever the recipient is using GPS, which could mean a reduction in the number of parcels left lying around or going undelivered. They are also working to get their 30-minute delivery service off the ground as well and this would mean that much of your shopping could be done in a fraction of the time it takes now, with your purchases delivered straight to your doorstep.

- **Disaster Management**

This could be one of the single most important uses for drones given the amount of unrestricted chaos we see after a disaster, not to mention the way that resources are misused and mismanaged. This is the same, no matter whether the disaster is a natural one or a manmade one and drones could of great help in these situations.

Utilizing powerful camera equipment, drones would be able to get clear images of destruction and debris in certain areas, providing a good picture and information about the effect of the disaster without the expense of full-size manned helicopters doing flyovers. And, once again, because of their size, drones would be able to get into harder to reach areas that a helicopter couldn't and get even more information and much closer views.

- **Healthcare/Rescue**

Rescue operations tend to be, more often than not, a real fight against time and there is no time for hanging about. Drones can be extremely useful here; with the addition of thermal sensors onboard, they can find missing persons and, at night, are even more effective, especially where the terrain is somewhat challenging to search and rescue teams on foot.

Drones can be put up into the air very quickly and get traverse through smaller spaces. Not only are they useful for search and rescue missions though, they can also be used for delivering medicines, food parcels, and other useful items to locations that are cut off through some kind of disaster, getting help where its needed before the rescue teams can get there

• Archaeological Surveying

Up until now, archeological surveys have cost a great deal in terms of energy and time but now, by using drones, these are much easier, quicker and can give us a much wider, clearer picture of an archeological site.

• Geographic Mapping

Drones have also proven to be useful tools in 3D mapping, especially in parts of the world that cannot easily be accessed any other way. This includes parts of the world's coastline that are particularly dangerous, mountain tops that can't otherwise be accessed and other tough-to-enter areas. Drones can get to these places and have provided more details, leading to more accurate mapping.

- **Law Enforcement**

Drones are a potential goldmine for law enforcement simply because they can be an eye in the sky in more places than law enforcement personnel can be at any one time. Because of their ability to hover on the fringes unobtrusive and unobserved by many people, drones can be used for surveillance purposes and for public safety.

Drones can easily monitor crowds, far better than personnel on the ground can, and they can easily spot if an emergency occurs or if criminal activity is happening. And they can also be used for monitoring crime scenes, providing a much clearer picture and more information. In cases of fire, drones could be sent in first to determine what the situation is before fire and rescue crews are sent in.

Lastly, another great use of drones is border patrol, where criminal activity like drug smuggling and human trafficking can be better detected.

• Safety Inspections

Some businesses have a need for carrying out safety inspections on a regular basis, such as power companies, gas pipelines, oil pipelines, bridges that are being constructed, and so on. Drones can easily do these inspections much faster than a human and can gather much more information. Aerial surveys can help to improve construction and performance on sites and, as the drones can get in much closer, they can provide much more detailed pictures of a construction site.

• Agricultural Industry

Drones are also being put to use in the agricultural industry, particularly for those farming on a large scale. Drones can provide very detailed analyses of large crop areas, tell a complete story of crop performance and easily identify any problem areas. Crop health can be studied using infrared sensors and give farmers an early warning of a problem that needs immediate attention, and this can all be done at a reduced cost and without any physical impact on the crop fields. Not only does this lead to better and healthier crops, it can also improve the yield of the crop.

- **Monitoring Wildlife**

In the same way that drones monitor crop fields, they can also be used to monitor fauna in some areas. This has a couple of advantages – firstly, it could eventually lead to cessation of poaching and secondly, it gives us a more unobtrusive way of studying wildlife behavior and patterns. The animals are not disturbed and are not affected in any adverse way and drones can also be used with thermal sensors at night, providing a complete picture.

• Weather Forecasts

This could potentially be one of the most revolutionary and ground-breaking uses of drones. Because of the quality of the cameras and the sensors on the drones they can be sent out to gather some very important information that could lead to much better, more accurate weather forecasting. They could be put up into a tornado, a hurricane, or other weather phenomena and send back useful images and data about the behavior and the patterns of these storms, leading to better warning systems in the future.

• Aerial Photography

Photography is the most common and popular use for drones and with the improvements in technology we have witnessed in the last few years, more drones than ever before can carry heavier cameras, providing better photography of certain areas. Not only are the cameras much better, the drones are far more stable and can provide images that are a good deal clearer.

With the addition of a gimbal, the cameras are more stable and with FPV or live streaming via Wi-Fi, you can see what the drone sees from a first-person point of view, not to mention being able to control them from a smartphone or tablet.

Creative Drone Uses

So, we've talked about some of the practical and useful drone uses but what about some of the more creative ones? Here are a few things that you could do with your drone today:

- ## Take a Selfie

Using a drone takes selfies to the next level and with more people than ever before taking selfies and sharing them everywhere, why not be a little different and use your drone? You can get more people or more landscape into your picture, you can take them from different angles and heights and you can control them from where you stand, making it very easy to produce selfies from a longer distance.

- ## Racing

Drone racing is becoming an incredibly popular activity now with race meets taking place everywhere. One of the more common activities is racing in the woods, providing a lot of fun and all the thrills you would expect. However, to do this, you do need a good agile drone and be a little more experienced at controlling it!

Drones are proving to be very useful in many different areas. No more are they limited to the military; more and more business sectors are now taking advantage of what a drone can do to boost their customer service and provide better services to more remote areas. While activities like racing are now popular, some people still use drones as toys, flying smaller ones indoors and larger ones just around their own premises. In the future though, you can expect to see drones come into more use in more business areas.

Lastly, we are going to look at community support and legalities of flying drones.

Chapter 7: Further Information

Now we come to the part about safety, legal issues and so on, all very relevant and a must-read section before you fly your drone anywhere other than in your own yard. There are lots of rules that pertain to drones and drone flying but, before you can find out what they are in your area, you need to know exactly how you are going to be using your drone. If it's just for fun then the rules are not so strict but if you are going to be making some money with your drone than there are things you must do, such as pass tests from the FAA and get certification. Once you have that certification you can then use your drone for aerial photography and videos that you are going to sell as stock images, for weddings, for real estate, for films, for all sorts. For the rest, if you just want to have a bit of fun, there are some different rules:

Registration

Until recently, if you were flying your drone for fun, it had to be registered with the FAA before it could be taken outdoors if it weighed 8 ounces or more. The cost for this was small, just $5 for a three-year registration on as many of these drones as you wanted. All you needed to do was pop a sticker on the drone that had an ID number on it. Now, however, the Federal Court of Appeals has overruled this, stating that the FAA was over the top with its requirements.

At the time of writing, there is no requirement for drones to be registered, however, as the case goes to the Court of Appeal, this may change. If you are using a DJI drone, you are required to register your drone and activate it using the official DJI control app, otherwise, it won't be fully functional. No other manufacturers have followed up on this yet.

Guidelines

Putting aside the issue of registration, there are still some guidelines that you have to follow, and those basic guidelines are:

- You cannot fly your drone above 400 feet
- You must keep full line of sight with your drone
- You must not fly anywhere near any other aircraft, in particular, airports
- You must not fly your drone over the top of people
- You must not fly your drone over the top of a stadium or any sports event
- You must not fly your drone over or near to disaster or emergency areas, like fires, as you could be seen as hampering rescue efforts
- You must not fly your drone whilst under the influence of alcohol or other substances
- Always be aware of any requirements regarding airspace in your area

If you are not sure how close to an airport you can or can't fly your drone, consider downloading and using an app like AirMap. This will tell you where you can and cannot fly. At the moment, the guidelines state that you must not be closer to an airport than 5 miles; if you are closer, you must notify the control tower

Most of these guidelines really are just common sense and that is something you are going to need in abundance when you fly your drone. As well as the rules issued by the FAA, you must also keep in mind that drones have also been banned from use within National Parks, something which is disappointing because these are places where you could get some stunning aerial photographs and video. However, it is only fair that some areas, including wildlife areas, should be able to be free of any technical disruption or distraction. It is also worth noting that you cannot fly in the airspace around Washington DC.

When Conflicts Arises

When you are flying your drone, especially in public areas, there is a chance that conflict of some kind will arise between you and another person or persons. There are those who are quite happy to take aim and shoot at drones with a gun so how do you deal with a situation like this, or where a person really doesn't want you flying your drone near them.

First things first; if a person does shoot at your drone or damage it in any other way then the first thing to do is notify the local police. However, if you, try to calm the situation down first. Be nice to the person, talk to them about what you are doing and about any concerns that they may have. Show them the images you are taking, the videos you are shooting; there are people who think that they are being spied on by a drone, so you can show them that that's not what you are doing.

Of course, you can only do this if you are given the chance and you will come across those who are not too reasonable. In these cases, be very aware of the position you are in. Are you in your own backyard? Are you on public ground? If either of these applies, it is your right to fly your drone. However, if you are standing on or flying over private property then things are different. The owner of the property is within their rights to tell you to bring your drone down and get off the premises. Always comply with those requests.

Be Wise

While it might be a great deal of fun to fly your drone and it's great that you can get images you would never have been able to get before, do be sensible about things.

Follow the FAA rules, don't get into conflict with other people and things will be more enjoyable. Obviously, you should not be flying a drone over a crowd of people and that really is just common sense.

You should also be careful about the time of day that you choose to fly your drone. Flying at sunrise means you are less likely to interfere with other people and you can get some cracking photographs as well.

Learn the rules, follow them to the letter, be sensible and your flying enjoyment will increase.

Connect with the Drone Community

To become proficient at flying your drone, you need to practice, you need to spend hours with your drone it. Not just flying it but crashing it, playing around with it, learning every part of it as well as learning everything you can about drones, flight and everything else that goes with them. Yes, your manual can give you some information but that is only the beginning. To truly get involved, you need to join the drone community.

There are millions of people the world over who are flying drones, building on the current technology and generally pushing things further than ever before and most of these can be found in communities on the internet. Here, you will find a wealth of information and support from like-minded people. But where do you start?

The most obvious place is to start at the website of the manufacturer of your particular drone. The more popular a particular make, the more bugs and issues will be found and sent back to the manufacturer. They then keep their websites updated with the latest news and this should the single most accurate point of information that you have.

If you have a DJO, the most popular model of drone, then you can head straight to their very own support pages. They have extensive support information and, if the answer to your question isn't there then they have a verity of contact methods where their technicians will try to help you.

Look online for large communities that are not sponsored or run by manufacturers. There are plenty of them and you will find plenty of help, along with ideas for what to do with your drone, tricks that you can try and other helpful information that you won't get from the manufacturer.

Discussion Forums

On occasion, you might find that the information on the manufacturer's website is either too technical for you or really doesn't answer your questions. If that's the case, look for a user forum for that manufacturer – there will sometimes be a link to one on their website.

A forum is a community of people online who get together to discuss their drones, problems, regulations, rules and other useful information. To join, you simply create a profile and post topics on the forum. Other users will then join in and give their advice, opinions, and support and this is one of the best ways to meet up with other drone enthusiasts. If you use a manufacturer's user forum, you stand a better chance of getting the help that you need.

However, although a forum is a fantastic way of getting answers if those answers do not come from those who with an authority on the subject then you should act with caution. An authority is someone who acts officially for the manufacturer of your drone and these are the only ones who can provide authoritative information so be careful – if you take all advice given as gospel, you could end up in a heap of trouble.

If you own a Parrot drone, then head to their website to look at their forums. This is an incredibly well-organized forum, with all information relating directly to the Parrot Drone. There are lots of distinct categories on the forum, pointing in exactly the right direction to get your answers. There are forum sections for each specific Parrot drone and then there are more general categories as well. Each category has specific topics in it, making it even simpler to get what you need and each of these is its own little discussion forum.

If the manufacturer of your drone doesn't provide a user forum or if that forum is not very comprehensive, then look elsewhere. There are lots of community forums where you can find what you want so just run a search on the internet and dig in.

Forum Rules

There are lots of people on a forum, many of whom have interesting views and love to talk to others but tend to keep themselves relatively anonymous. Forum users are a fussy lot and are quite happy to tell you if you deviate from the forum rules and yes, they do have them, some having an extensive list of rules to follow. Before you dive in and start posting in any forum, make sure you read and familiarize yourself with the rules, some of which are fairly obvious, others not so much. While each forum will have its own set of rules, these are some of the more common ones:

- Don't use rude language
- Don't be derogatory or defame anyone
- Don't use the forum as a place for advertising
- Don't be negative about the drone manufacturer
- Don't encourage law-breaking

Most forums are under the control of a moderator and these work on behalf of the owner, making sure that all the rules are stuck to. As well, they will attempt to sort out any issues between users and will also be on hand for feedback and expert advice. Learn the rules and you will stay under their radar and also bear in mind that many forum users will also police the forums.

Before you post, keep the following in mind:

- Read the rules and make sure you fully understand them. This will stop you getting into trouble because, even if the moderator doesn't pick up on something, no doubt another user will!
- Don't jump in too quickly with opinions or new posts. Things move very quickly in the drone world and many people don't have the patience to wait and see. Always look thoroughly through a forum for an answer before you post – duplicate questions, upset other users
- Watch out for forum trolls. These may be a great place to talk about drone issues with people in the know but there are also a lot of users who like nothing more than to drown the

forum in useless arguments and debates, doing nothing more than dragging it down. Just be aware that there may be a troublemaker or two lurking in the background and don't get drawn into pointless discussions.

Conclusion

Drones are far more accessible by just about anybody now than they have ever been, with millions being sold every year. These are not just toys; while some may be small and light, suitable only for flying around the house or just for practicing, others are expensive, sophisticated pieces of kit that can be used in so many ways – from general fun, racing and shooting video to more helpful ways like deliveries, policing the neighborhood and so on. For the most part, yours will be a fun hobby, at least to start with while you gain confidence and skill in flying and controlling them.

No matter what you choose to use your drone for if this is your first one you need to be prepared to put in the time in learning your drone. Learn how it's made, how it flies and how to control it. Practice, practice and then practice some more. I guarantee you, once you have the basics, once you know your drone inside out, you will want to move on to bigger and better ones, drones that can go further, go where you want them to and even follow you around.

In my next book, I will be talking you through how to build your own drone but, for now, take advantage of the vast array of different ones on the market and start having some serious fun!

Dear Reader,

I would love to hear your opinion about my book. In the world of book publishing, there are few things more valuable than honest reviews from attentive readers.

Your review will help other readers find out if this book is for them. It will also help me reach more readers by increasing the visibility of my book.

Thank you!

References

https://time.com

https://www.dronezon.com

https://droneandquadcopter.com

http://smashingdrones.com

http://droneguru.net

https://dummies.com

https://pcmag.com

https://www.uavsystemsinternational.com

www.ingramcontent.com/pod-product-compliance
Lightning Source LLC
LaVergne TN
LVHW010236200726
843506LV00014B/3005